Daniel Luiz Pereira

Evaluation of cholinesterase in the saliva of tobacco farmers

Daniel Luiz Pereira

Evaluation of cholinesterase in the saliva of tobacco farmers

A new method for detecting pesticide poisoning

ScienciaScripts

Imprint

Cover image: Provided by the author

This book is a translation from the original published under ISBN 978-613-9-72229-7.

Publisher:
Sciencia Scripts
is a trademark of
Dodo Books Indian Ocean Ltd. and OmniScriptum S.R.L publishing group

120 High Road, East Finchley, London, N2 9ED, United Kingdom
Str. Armeneasca 28/1, office 1, Chisinau MD-2012, Republic of Moldova, Europe
Printed at: see last page
ISBN: 978-620-7-73620-1

"There are men who fight one day and are good.
There are others who fight for a year and are better.
There are those who fight for many years and are very good.
But there are those who fight all their lives.
Those are the essentials."
Bertolt Brecht.

DEDICATORY

To my family who have always been there for me at all times...

To my wife and daughter who have always been by my side...

To the teachers for the knowledge passed on during the 4.5 years of the course...

ACKNOWLEDGEMENTS

To Prof. Msc. Eduardo Ottobelli Chielle, supervisor, who accompanied every stage of this work.
To the other teachers and fellow students with whom I have spent these 4.5 years and who have given me exchanges of knowledge and experiences that I will carry with me forever.
To the professionals interviewed, who provided the information needed to carry out this study.
To my family for their support and understanding.

SUMMARY

SUMMARY

Introduction: The World Health Organisation estimates that around three million people suffer from acute pesticide poisoning, and among the classes of workers most exposed to these agents are tobacco farmers. Organophosphate poisoning causes inhibition of cholinesterase enzyme activity, so determining cholinesterase activity acts as a biological indicator of the effect of exposure to organophosphate insecticides. **Aim:** This study evaluated the activity of the enzyme cholinesterase in saliva samples from tobacco growers, in order to check whether or not there was occupational intoxication due to excessive exposure to pesticides during the tobacco growing period and whether this biological sample would be able to provide this information.

Results: When analysing the results of the collection of the 23 volunteers studied, it was possible to see that both sexes are in contact with pesticides, where 73.91% said they had already had direct contact with the pesticide on their skin, eyes or another part of their body, this is related to the fact that 60.8% of the farmers interviewed do not make proper use of the safety equipment indicated. With regard to the symptoms felt after handling the products, all said they had experienced at least one symptom. With regard to the results obtained from the expression of cholinesterase enzyme activity in saliva samples, there was a significant reduction ($P<0.0001$) in the second collection when compared to the first, before exposure, highlighting the contact of farmers with pesticides, the same was observed with creatinine, while when evaluating AST and ALT, no significant differences were observed between the collections.

Conclusion: In this study, one of the main points that was identified was farmers' disregard for the use of PPE for their safety and health during the preparation, handling and application of pesticides, as this is due to a lack of information about the importance of these habits for their health. This exacerbated exposure leads to a decrease in cholinesterase, an enzyme that hydrolyses acetylcholine and regulates the transmission of nerve impulses in the nerve synapse and neuromuscular junction. Saliva as a biological sample for this test proved to be reliable as its results were in line with those expected.

1 INTRODUCTION

In order to meet the need to increase agricultural production, it has become necessary to develop new technologies, ranging from soil preparation to harvesting and product processing. Among these technologies is the application of pesticides, a technique that has caused a great deal of environmental and health damage due to its toxicity (TROIAN et al, 2009).

This contamination is present in food, animals and humans. It is therefore of the utmost importance that, when using these products, people are aware of their choice, mixture, application and expiry date. The indiscriminate use of pesticides on crops has caused major economic, human health and environmental damage.

The World Health Organisation estimates that around three million people suffer from acute pesticide poisoning, and that the number of deaths per year reaches 220,000. Of these, around 70 per cent occur in developing countries. The workers affected include those who have direct or indirect contact with these products, as well as consumers of food contaminated with these substances (TROIAN et al, 2009).

According to Stroparo et al (2011), one of the classes of farmers who are directly exposed to these agents are tobacco growers, who are in daily contact with several of these compounds. According to the Pan American Health Organisation, among those most exposed to pesticide poisoning are the workers who apply the product, those who dilute or prepare the "syrups" and those who enter the fields after application. Tobacco growers in general are the people who are in direct and continuous contact with all the contamination processes.

Organophosphate insecticides, which are the most responsible for poisoning and deaths in the country, and carbamates, which are inhibitors of cholinesterases (fundamental enzymes of the nervous system), can be absorbed by the body through skin contact, ingestion or inhalation. These pesticides act on the central nervous system, the blood and other organs (STROPARO et al, 2011).

Cholinesterase is an enzyme that acts in the human body as a chemical mediator. They are inhibited in the presence of organophosphate and carbamate class pesticides. When exposed to these compounds, the activity of cholinesterase decreases quantitatively and proportionally to the intensity and time of exposure, so these mediators are widely used in research into poisoning by these pesticides (STROPARO et al, 2011).

Organophosphate poisoning causes inhibition of the activity of the cholinesterase enzyme, i.e. determination of blood cholinesterase activity acts as a biological indicator of the effect of exposure to organophosphate insecticides (MIDIO et al, 1995).

With the inhibition of cholinesterase, there is an accumulation of the neurotransmitter acetylcholine, leading to the induction of three types of clinical signs: muscarinic, characterised by sweating, sialorrhoea, gastrointestinal hypermotility, diarrhoea and bradycardia; nicotinic effects, which result from the action of acetylcholine on the motor plates, causing muscle rigidity, tremors, paresis and paralysis, and also in the central nervous system, leading to restlessness, ataxia, convulsion, depression and coma (CAZARIN, 2007).

Expectations are that saliva tests could soon be used to diagnose breast cancer, viral and bacterial diseases, the presence of hormones, drugs and toxins, and for nutritional deficiencies (CHESTER, 2005).

Given the importance of this subject, the aim of this study was to assess the activity of the enzyme cholinesterase in saliva samples from tobacco farmers in order to determine whether or not there was occupational poisoning due to excessive exposure to pesticides during the tobacco-growing period. It should be noted that research into the activity of cholinesterase in saliva is of the utmost importance in order to discover analyses in different biological samples that facilitate the collection, processing and diagnosis of pesticide poisoning.

2 OBJECTIVES

2.1 GENERAL OBJECTIVE

To assess the activity of the enzyme cholinesterase in saliva samples from tobacco growers, in order to verify whether or not there was occupational intoxication due to excessive exposure to pesticides during the tobacco-growing period.

2.1.2 SPECIFIC OBJECTIVES

Research into whether saliva is an effective biological sample for checking pesticide poisoning.

To correlate the symptoms reported by tobacco farmers with the expression of salivary cholinesterase.

To verify the use of personal protective equipment with salivary cholinesterase values and the levels of intoxication with the non-use of personal protective equipment (PPE).

Assess whether farmers have ever come into contact with pesticides directly on their skin, in their eyes or elsewhere,

Assess renal and hepatic function markers.

Checking the population's exposure time to pesticides.

3 THEORETICAL FRAMEWORK

3.1 GENERAL CONSIDERATIONS

The use of pesticides in agriculture began in the 1920s, a time when the toxicological effects of these products were unknown. Its expansion began after the Second World War, when it was used as a chemical weapon (RIBEIRO et al, 2007).

The application of these products in Brazil was first used in public health programmes to combat vectors and control parasites. Since the 1960s, it has been used more intensively in agriculture.

The extensive use of various types of pesticides today has created a serious public health problem in developing countries, especially those with agriculture as their economic base. Brazil is one of the four largest consumers of pesticides in the world and the second largest producer of tobacco, with production concentrated in the three southern states of Paraná, Santa Catarina and Rio Grande do Sul. This has led to reports of occupational poisoning and deaths related to exposure to pesticides (TROIAN et al, 2009).

3.2 AGROTOXICS

Expressions such as pesticide, pesticide, biocide, phytosanitary, pesticide, poison and remedy are the various names given to the same group of chemical substances, which have been called 'agrochemicals', and have been defined since Decree No. 4,074 of 4 January 2002, which regulated Law No. 7,802/1989, as: (...) products and physical, chemical or biological processes intended for use in the production, storage and processing of agricultural products, in pastures, in the protection of native or planted forests, and other ecosystems.) products and agents of physical, chemical or biological processes, intended for use in the production sectors, in the storage and processing of agricultural products, in pastures, in the protection of native or planted forests, and other ecosystems and urban, water and industrial environments, whose purpose is to alter the composition of flora or fauna, in order to preserve them from the harmful action of living beings considered harmful, as well as the substances of products used as defoliants, desiccants, growth stimulators and inhibitors (TROIAN et al, 2009).

According to Ribeiro et al (2007) these substances are used in agriculture, but in some cases they can be used in public health to eliminate and control vectors that transmit endemic diseases such as Chagas disease, malaria and dengue.

The use of agrochemicals in agriculture aims to preserve food from the harmful actions of living beings considered harmful, to increase the production and quality of food and to reduce labour in the fields (PAHO/WHO, 1996).

The importance of using agrochemicals in the final result of agricultural productivity cannot be overlooked, as it provides higher quality production at a lower cost. However, it is important to emphasise the dangers this poses to those who are exposed to these agents and to those who consume the products containing the pesticides (RIBEIRO et al, 2007).

According to SINITOX (Sistema Nacional de Intoxicação por Agrotóxicos) (2009), human poisoning by pesticides in Brazil increased gradually between 1985 and 2007.

The risks of poisoning are great for those who have intense contact with concentrated pesticides. The skin is the organ most exposed during spraying. Workers can come into contact with pesticides in a number of situations, including during transport, storage, preparation of syrups, handling and cleaning of spraying equipment and during the disposal of empty containers (RIBEIRO et al, 2007).

There are two groups of pesticides that are highly toxic to humans and animals because they are cholinesterase inhibitors: organophosphates and carbamates. Organophosphates, chemical compounds that have a phosphorus atom in their structure, are widely used in farming and have caused accidental poisoning in animals and humans, as well as suicide attempts (RIBEIRO et al, 2007).

With the National Development Plan (PND) created in 1975, which was responsible for opening up trade in pesticides in the country, farmers were obliged to buy these chemical substances with rural credit resources, by designating a set quota of pesticides for each loan applied for. Combined with this obligation and the manufacturers' advertising, there has been a gradual increase in the use of pesticides in Brazil. The result is that we have become one of the countries that consume the most pesticides, resulting in numerous problems for both people's health and the environment (TROIAN et al, 2009).

More than 1.5 million tonnes of pesticides are manufactured worldwide every year (SATOH, 2006). The considerable increase in the use of agrochemicals in Brazil in recent years has placed it at the top of the world rankings (PRONAF, 2010; REBELE et al., 2010; SINDAG, 2011).

According to Troian et al. (2009), Brazil is the eighth largest consumer of pesticides per hectare in the world, with herbicides (a substance used to destroy pests that prevent the desired crop from growing) and insecticides (a product used to kill insects) accounting for 60% of the products sold in the country. In the state of Mato Grosso do Sul, between 1992 and 2002, there were 1,355 reports of poisoning caused by handling pesticides in agriculture, of which around 500 were due to voluntary ingestion. It is assumed that the highest numbers of poisoning cases are concentrated in towns with smaller farms, where farmers have more manual contact with pesticides.

3.3 ORGANOPHOSPHATE AND CARBAMATE INSECTICIDES

Organophosphates were first synthesised at the beginning of the 19th century in order to investigate their effects as insecticides. It was later used as a weapon of war. In 1941, during the Second World War, they were reintroduced for agricultural use (SATOH, 2006).

Organophosphate pesticides are esters, amides or derivatives of phosphoric or phosphorous acid, in which all the hydrogen atoms are replaced by organic forms (PASIANI, 2012).

These insecticides are stable when stored in low-temperature, dark environments. However, when exposed to heat, light or water, chemical changes can occur, mainly hydrolysis and oxidation (CHAMBERS et al., 2001).

These chemical reactions involving organophosphates can result in a more or less toxic product. When the complete hydrolysis reaction occurs, organophosphates are detoxified, resulting in less reactive products. The non-enzymatic oxidation reaction can occur in products stored at high temperatures or exposed to light (WHO, 1986; CHAMBERS et al., 2001).

In the mid-1950s, with research into insecticidal properties that were less toxic to mammals than organophosphates, carbomate insecticides - carbamic acid esters - returned to activity as herbicides and fungicides (PASIANI, 2012).

Carbamic acid can be stabilised by forming a salt or by synthesising alkyl esters. Substituting hydrogen for a methyl group on the nitrogen atom results in the formation of N-monomethylcarbamic acid, which combined with an aryl ester substituent introduces insecticidal properties. Another group of carbamate insecticides are the aliphatic oxime derivatives, which are called methylcarbamoyloxy or oxime methylcarbamates and have a high degree of toxicity. There are 52 organophosphate and carbamate products registered in Brazil (PASIANI, 2012).

3.4 THE USE OF PESTICIDES AND TOBACCO CULTIVATION

Troian et al (2009) reports that pesticides are widely used in large-scale monoculture systems, such as soya, sugar cane, maize, coffee, citrus fruits, rice and cotton. However, some less significant crops, such as tobacco, grapes, strawberries, potatoes, tomatoes and other vegetables and fruit, also use large quantities of pesticides.

Among the crops most associated with the occurrence of poisoning in Brazil, tobacco stood out, with 127 cases recorded. This is due to the fact that most producers don't use PPE. Not using this equipment exposes producers to toxic agents, the liver and kidneys being the most affected (ALMEIDA, 2011).

Brazil is the largest exporter of leaf tobacco, with most of its production concentrated in the southern region of the country, where around 90 per cent of the national area is located. Although tobacco is considered a summer crop, its activities are spread out over almost the entire year. Preparations for the tobacco seedlings begin in early spring. The end of the harvest, i.e. harvesting, drying and grading takes place in mid-summer, after which the farmer starts preparing the land for the next harvest (TROIAN et al, 2009).

When growing tobacco, farmers come into direct contact with pesticides and the temperatures are higher. As a result, the use of safety equipment is neglected by farmers who are exposed to health risks. Personal protective equipment, known by the acronym PPE, is defined by Regulatory Standard number 6, approved by Ministerial Order no. 3.214/78 of the Ministry of Labour, as any device or product, for individual use by the worker, designed to protect against risks likely to threaten safety and health at work (TROIAN et al, 2009).

According to Almeida et al (2011), when farmers use pesticides indiscriminately, abusively and incorrectly, they become co-responsible for the imminent damage to their own health, the contamination of food and soil, causing damage to the environment and the beings that live in it.

Pesticides are subdivided into 23 categories, including insecticides, acaricides, fungicides and herbicides. These can be classified into four toxicological classes: class I, extremely toxic; class II, highly toxic; class III, moderately toxic; class IV, low toxicity, indicated on product packaging with red, yellow, blue and green bands respectively (ANVISA, 2011).

Various pesticides are recommended in the tobacco production chain, from seedbeds to crops, including Manzate 800, Orthene 750 BR and Gamit (TROIAN et al, 2009).

3.5 INTOXICATION

According to the International Organisation of Consumers' Unions, every 4 hours an agricultural worker in developing countries dies from pesticide poisoning. The group of products involved in most poisonings are organophosphates (RIBEIRO et al, 2007).

According to Ribeiro et al (2007) organophosphate and carbamate insecticides are powerful inhibitors of enzymes that are fundamental to the functioning of the nervous system. When they come into contact with the skin, through ingestion or inhalation, these substances are absorbed by the human body, causing damage to the central nervous system, the blood and other organs.

According to Trápe (1999), pesticides act by inhibiting enzymes in the body called cholinesterase. These enzymes are present in the transmission of nerve impulses in various organs and muscles of the body. When contamination occurs, there is a link between the enzymes and the poison, preventing them from carrying out their function, resulting in a series of symptoms and clinical signs, including cholinergic syndrome: sweating, excessive salivation, pinpoint pupils (miosis), bronchial hypersecretion, vomiting, cramps and diarrhoea; nicotinic syndrome: tremors, muscle tremors, changes in blood pressure; and the neurological syndrome: mental confusion, difficulty walking, convulsions, cardio-respiratory depression, coma and death.

Peretrode pesticides, on the other hand, are absorbed by the digestive tract, respiratory tract and skin. They are generally not very toxic, but they irritate the eyes and mucous membranes, causing skin allergies and bronchial asthma (RIBEIRO et al, 2007).

Once absorbed into the body, these chemical substances pass into the bloodstream and are distributed throughout the body. When they reach certain organs, they are biotransformed, producing toxic effects and then eliminated by the body through urine (RIBEIRO et al, 2007).

There are three types of poisoning that are usually caused by pesticides. Acute poisoning is when signs appear soon after contact with highly toxic products, but it can be mild, moderate or severe, depending on the amount of toxic substance absorbed. The signs and symptoms are clear and objective, which facilitates the diagnosis of this pathological form. Intoxication described as subacute occurs as a result of moderate or minor exposure to highly toxic products, and has a slower symptomatic evolution, in which case the symptoms are usually subjective and vague, such as: headache, weakness, malaise, stomach pain and drowsiness, among others. However, as these symptoms are considered non-specific, they do not usually motivate the patient to seek emergency care and are often managed without medical assessment (ALMEIDA, 2002).

Known as chronic intoxication, it is characterised by the late onset, in months or years, of small or moderate exposure to toxic products or multiple products, causing irreversible damage such as paralysis and neoplasms (Almeida 2002).

Chronic exposure to organophosphates can lead to late neurological effects, known as late peripheral neurotoxicity or intermediate syndrome, which appears after the acute cholinergic effects have recovered (LOTTI, 2001). The first symptoms are muscle pain, followed by tingling, burning and weakness in the legs, with a reduction in reflexes. In more severe intoxications, the upper limbs can be affected, leading to quadriplegia (PASIANI, 2012).

Organophosphate insecticides can be absorbed through the skin, by ingestion or inhalation, and are substances that are stored in adipose tissue, cross the blood-brain and placental barriers and are generally metabolised in the liver. In cases of death due to organophosphate insecticide poisoning, the cause consists primarily of respiratory failure, which is usually accompanied by a secondary cardiovascular component (ALMEIDA, 2002).

According to Ribeiro et al (2007) there are associations between acute intoxication and the use of psychiatric drugs, known as neuropsychological sequelae. Long-term exposure to organophosphates leaves neurobehavioural sequelae that can evolve into depression, all of which adds up to a series of economic and social problems. The damage caused to the brain by organophosphates can leave sequelae such as apathy, irritability, loss of concentration and memory; problems that can evolve into depression and even suicide.

In Brazil, there are approximately 32 Poison Control Centres (CCIs) located in 17 states, which provide care for acute poisoning or those in the process of exacerbating the chronic phenomenon in workers. Statistical data from the Toxicology Centres in Belo Horizonte, Campinas, Florianópolis, Ribeirão Preto, Londrina and Maringá show that, of 495 cases of occupational poisoning, around 34.9% were due to organophosphates; of 622 cases of attempted suicide, 38.1% resulted from the use of compounds from this group; of 38 cases of death, 44.7% were due to organophosphates (RIBEIRO et al, 2007).

3.5.1 **Laboratory diagnosis of intoxication**

The diagnosis is made through anamnesis, the farmer's history of poisoning and the clinical manifestations presented, followed by the evaluation of biomarkers such as cholinesterase, which will show alterations depending on the magnitude of exposure (FERNANDES, 2010).

Laboratory control of occupational exposure is usually carried out by determining the cholinesterase activity in workers' blood. When cases are extremely serious, it is important to carry out other tests such as blood counts, chest X-rays, renal function, blood gases, glycaemia and liver function (FARIA et al, 2009: STROPARO, 2011).

3.5.2 **Biomarkers of exposure to pesticides.**

Constant biological monitoring serves to measure biomarkers in tissues and body fluids in order to assess the exposure and health risk of a group of substances or one in particular, compared to the appropriate reference (WHO, 1986).

According to Who (1986) and Amorin (2003), these biomarkers can be classified into three types:

- Exposure: is an exogenous substance, its metabolite or the product of interaction between the xenobiotic agent and some target molecule or cell that is measured in the organism;

- Effect: a biological parameter that reflects the interaction with an exogenous substance and can be associated with impaired health or disease;
- Suceptibility: is an indicator of an organism's inherent or acquired ability to respond to exposure to a specific substance.

According to Who (1993), biomarkers are used to diagnose poisoning and assess the effectiveness of treatment. In an occupational context, they provide a complementary means of reviewing the adequacy of protective measures, including working practices and conditions.

The Ministry of Labour and Employment issued NR-7, which establishes biomarkers of occupational exposure to 26 chemical agents, including organophosphate and carbamate insecticides. An inhibition of more than 30% for acetylcholinesterase (AChE) indicates excessive exposure and this can be associated with adverse effects; in the case of butyrylcholinesterase (BuChE), if it is depressed by 50% or more, it indicates excessive exposure (PASIANI, 2012).

According to Pasiani (2012), in order to assess the level of exposure to organophosphates and carbamates, one can compare the average enzyme activities of a study group in the period of exposure and non-exposure, or calculate the cut-off point according to the results found in the analyses with the results found in the analyses of the enzymes of the non-exposed population.

As the inter-individual variation of biomarkers is large compared to intra-individual variation, analyses of paired samples (before and after exposure) can considerably increase the power of biomarkers to detect exposure. The inter-individual variation in AChE and BuChE activity is 10%-18% and 15%-25% respectively, and the intra-individual variation is around 3%-7% and around 6% respectively (WHO, 1986; MARONI et al., 2000).

3.6 COLINESTERASE

Cholinesterase is the group of enzymes known as esterases, they are polymeric and occur in multiple molecular forms. Erythrocyte or true cholinesterase (AChE) *is* found in the autonomic nerves, somatic motor nerves and the erythrocyte membrane. Plasma cholinesterase or pseudocholinesterase (BChE) is found in the liver, intestine, brain and plasma (FUKUTO, 1990; RANG et al., 2004).

The groups of hydrolase enzymes responsible for acting as chemical mediators by catalysing the hydrolysis of choline esters are what we call cholinesterase. Many pesticides have the power to inhibit these enzymes, including organophosphate insecticides, which inhibit both plasma and erythrocyte cholinesterase (LARINI, 1999).

When organophosphates or carbamates act on acetylcholinesterase, this enzyme is no longer able to hydrolyse acetylcholine. The inhibition of acetylcholinesterase by organophosphates takes place through a chemical reaction, where the hydroxylated serine molecule in the enzyme's active site is phosphorised in a manner analogous to the uptake of acetylcholinesterase, forming an enzyme-inhibitor complex with the subsequent phosphorisation of the serine hydroxyl and the loss of the organophosphate leaving group. In some cases, this inhibition can become irreversible (PASIANI, 2012).

With carbamates, practically the same process takes place, the difference being that phosphorylation is replaced by carbamylation, which also results in inhibition of the enzyme. This inhibition is unstable, short-lived and reversible (REINER & SIMEON - RUDOLF, 2006).

AChE (acetylcholinesterase) inhibitors act as a pseudo-substrate by binding to the serine amino acid of the enzyme's active site, causing an increase in inhibition and thus an accumulation of acetylcholine in the synapses (FERNANDES et al; 2010, ARAÚJO et al; 2007).

Cholinesterase activity comes from the action of two enzymes, one in the erythrocyte membrane (erythrocyte cholinesterase or acetyl cholinesterase -AchE) and the other in the serum (plasma cholinesterase or butyryl cholinesterase -BuChE). The decrease in plasma cholinesterase content can remain for thirty days and in red blood cells for ninety days after the last contact with organic phosphates. This difference has proved to be a skilful way of differentiating poisoning, with AchE for poisoning suffered more recently and BuChE for more recent poisoning (RIBEIRO et al, 2007).

Unlike AChE, BuChE is not associated with cholinergic synapses and exhibits low substrate specificity. Its biological role is also not as well defined as AChE, but both are inhibited by organophosphates or carbamates (FUKUTO, 1990; THOMPSON, 1999).

In the diagnosis of poisoning by phosphorus insecticides, one of the best forms of prevention is monitoring with the determination of cholinesterase activity, where the level of the enzyme acetylcholinesterase in the blood is checked. The pseudocholinesterase found in serum decreases before that found in red blood cells and is therefore a biological indicator of acute exposure to organophosphate insecticides. The diagnosis of acute and severe anti-ChE poisoning is easily established on the basis of the history of exposure and characteristic signs and symptoms. In suspected cases of milder acute or chronic intoxication, the determination of ChE activities in erythrocytes and plasma usually establishes the diagnosis (RIBEIRO et al, 2007).

3.7 BIOCHEMICAL MARKERS

According to Goethel et al (2013), after being absorbed and distributed in the body, organophosphates are metabolised in the liver by a series of enzymes, undergoing chemical transformations. This favours an increase in the solubility of the pesticide, which facilitates its excretion through the kidneys. It is therefore possible that there are alterations in liver and kidney metabolism.

AST (aspartate aminotranferase) and ALT (alinine aminotranferase) are the most commonly used liver markers. In the case of renal markers, urea and creatinine are the most commonly used, in which alterations indicate renal insufficiency, which in many cases is associated with organophosphate poisoning, and there are already some studies that mainly prove the elevation of serum creatinine (FERNANDES et al, 2010; GOETHEL et al, 2013).

3.7.1 **AST and ALT**

Considered the largest organ in the human body, the liver weighs an average of 1.5kg. It is a highly complex organ responsible for carrying out various vital functions, some of which cannot be replaced even by the most modern therapeutic technologies (NUNES, 2007).

According to Barbosa (2009), the liver is an organ that is involved in various metabolic activities and by assessing the enzymes of certain pathways, its metabolic status can be predicted. Hepatic transaminases can change their activity depending on the brightness of the environment. They correspond to an important group of enzymes that catalyse the transfer of an amino group from an amino acid to a ketoacid.

Aminotransferases are essential enzymes involved in the central metabolism of the whole organism, very useful in assessing liver and heart function.

According to Pinto (2010), AST exists in both mitochondrial and cytoplasmic form, and is found mainly in the liver, heart, skeletal muscles and kidney. ALT is exclusively cytoplasmic and is found mainly in the liver and kidneys. Both are aminotransferases of clinical interest, however AST is much less specific for liver damage than ALT, this can be seen in analyses of patients with liver damage where ALT is always higher than AST. Only in alcoholic liver damage is the rise in AST about 2-3 times higher than the rise in ALT, because alcohol has an inhibitory effect on ALT synthesis.

These enzymes are important in the diagnosis of liver and heart damage caused by myocardial infarction, toxic drugs and infections, as in many situations they leak out of damaged cells and into the bloodstream. Determinations of these enzymes help to assess whether individuals who have been exposed to carbon tetrachloride, chloroform or other industrial solvents have liver damage, since this can cause liver degeneration, releasing various enzymes into the bloodstream (PINTO, 2010).

3.7.2 **Urea and Creatinine**

The kidneys are responsible for various functions, including filtration, reabsorption, homeostasis, endocrinological and metabolic functions. However, its primary function is to maintain homeostasis, regulating the internal environment predominantly by reabsorbing substances and ions filtered in the glomeruli and excreting other substances.

According to Sodré et al (2007), urea is the main nitrogen metabolite derived from the breakdown of proteins in the body, 90% of which is excreted by the kidneys and corresponds to approximately 75% of the non-protein nitrogen excreted. According to Klein et al (2008), serum urea is a parameter used to assess kidney function, but increases in urea levels are not always due to a decrease in glomerular filtration rate.

The main clinical usefulness of urea seems to lie in its determination in conjunction with creatinine. The serum urea/creatinine ratio can indicate different pathological states. A lower than expected value can be found in pathologies such as acute tubular necrosis, low protein intake, conditions of food deprivation or reduced urea synthesis due to liver failure. This high ratio can be analysed dichotomously with creatinine within the reference value, indicating processes that lead to reduced renal blood flow, increased protein intake or gastrointestinal bleeding; and with creatinine above the normal value, denoting post-renal obstructive processes such as tumours or urinary tract stenosis. Another use for urea is in its urinary dosage, which can provide crucial information in the field of nutrition and has been used in hospitalised patients to monitor special diets (SODRÉ et al, 2007).

Creatinine is a waste product of creatine. According to Sodré et al (2007), the transformation of creatine into creatinine takes place in muscle tissue, in which 1%-2% of free creatine is spontaneously and irreversibly converted into creatinine every day, it is freely filtered in the glomerulus, actively secreted in a small portion, but enough to overestimate the GFR (glomerular filtration rate).

Due to their simple and low-cost methods, urea and creatinine tests are widely used in laboratory and clinical routine (SANTOS; WAGNER, 2013). Although they are routinely used in clinical assessments, the sensitivity and specificity of urea and creatinine are debatable, since both do not provide data immediately after the onset of renal dysfunction, which compromises the choice of the ideal time for their assessment.

3.8 TREATMENT OF ORGANOPHOSPHATE POISONING

According to Ribeiro et al (2007), the treatment is specific and effective. Atropine, in sufficient doses, effectively antagonises local actions on muscarinic receptors, including increased tracheobronchial and salivary secretions, bronchoconstriction, bradycardia and, to a moderate degree, central and peripheral ganglionic actions. Larger doses are needed to obtain appreciable concentrations of atropine in the CNS. Atropine has virtually no peripheral neuromuscular effect or impairment. The latter action of anti-ChE agents, as well as all the other peripheral effects, can be reversed by pralidoxime, a cholinesterase reactivator.

In cases where extensive fat distribution has already occurred, toxicity will persist and symptoms may continue after the first treatment. In these cases it is necessary to continue treatment for several weeks (RIBEIRO et al, 2007).

When treating poisoning by these insecticides, the use of ContrathionTM (Pralidoxime) is indicated as a chemical antidote. In order to decontaminate the skin, the necessary measures must be taken at the same time to ensure the life of the victim, such as resuscitation and administration of antidotes (RIBEIRO et al, 2007).

According to Ribeiro et al (2007), to remove the chemical from the skin and hair, the patient should be washed with water, soap and shampoo. When the symptoms are weakness, ataxia and neurological problems, all contaminated clothing should always be removed as quickly as possible in order to remove the source of dermal exposure, and the patient, in a lying position, should be washed thoroughly. Chemical products in the eyes should be washed out with plenty of water for around 10 to 15 minutes.

Aid workers should not come into contact with contaminated clothing or the patient's vomit. Rubber gloves must be worn when washing the patient's skin and hair (RIBEIRO et al, 2007).

3.9 SALIVA AS A SAMPLE FOR A DIAGNOSTIC METHOD

Saliva is defined as a fluid of glandular origin that covers the surfaces of the mouth and has very varied physical and chemical properties. Its secretion is induced by psychological, mechanical, physical, chemical and biological stimuli, and the stimulated salivary flow can vary from 1.0 to 1.5 ml per minute, which facilitates the collection required for analysis. Its composition is also very different, containing ions of calcium, phosphate, bicarbonate, fluride, sodium and potassium, organic substances such as glycoproteins, digestive enzymes, glucose and urea. In the last 10 years, the use of saliva as a diagnostic method has advanced considerably, both to monitor oral and systemic diseases (SCHUTZEMBERGER et al, 2007).

There are numerous benefits to using saliva in tests to replace blood or urine. As it is painless to collect, it causes less anxiety. For clinicians and lab technicians, saliva tests don't expose them to the risk of contamination. Another important feature of saliva tests is that they are cheaper to store and transport than blood and urine tests (CHESTER, 2005).

On the other hand, for a salivary test to be used, it is necessary to have a reliable correlation between the levels of designated substances in saliva and in blood plasma. Simply having a little blood in the saliva makes the test inaccurate (CHESTER, 2005).

4 MATERIALS AND METHODS

4.1 STUDY DESIGN

The study was carried out with adult tobacco farmers over the age of 18, recruited in the municipality of Paraíso - SC, between August 2014 and May 2015. It was characterised as a prospective study which took into account the work process, the substances used, the safety practices adopted, the time of exposure and use of the products and reports of pathologies and intoxications possibly related to the use of pesticides in tobacco production.

For the toxicological monitoring of individuals occupationally exposed to organophosphate and carbamate pesticides, the Ministry of Labour recommends through NR 7 of Ordinance No. 3,214 of 08/06/1978 that plasma and/or erythrocyte cholinesterase levels be measured, determining pre-occupational and post-occupational enzyme activity. If the enzyme activity is inhibited by 30% or more compared to the pre-occupational period, the individual is considered intoxicated (NR 7 - Occupational Health Medical Control Programme; 2014).

4.1.1 **Population and sample**

The study population was made up of individual tobacco farmers from the municipality of Paraíso - SC, characterised as follows:

- First collection: 23 adult individuals over the age of 18, male or female, who carry out activities related to tobacco planting, cultivation, cleaning, sorting and pesticide application in 2014.
- Second collection**: The** same 23 individuals were collected as in the first collection.

4.1.2 **Intervention protocol**

The intervention protocol was submitted for review and approval by the Human Research Ethics Committee of the Universidade do Oeste de Santa Catarina - UNOESC, and was approved under protocol number 1.025.353 (attached). All the selected participants signed an informed consent form (APPENDIX A), after which each participant was given information about the objectives, risks and benefits of the study.

4.1.3 **Application of the questionnaire**

After signing the Free and Informed Consent Form, a questionnaire (Appendix A) with open and closed questions was administered to all the participants regarding the management and planting of tobacco.

4.1.4 **Laboratory analyses of saliva samples**

In order to carry out the laboratory analyses on saliva samples, it was necessary to rinse the mouth with water using light mouthwashes before collection. It is not recommended that samples be taken from cases of oral lesions with active or potential bleeding. To prevent gingival bleeding, subjects were advised to brush their teeth thoroughly at least two hours before collection. They then removed the top cap from the tube, placed the cotton wool in the suspended container under their tongue and waited 2-3 minutes. Some chewed lightly on the cotton wool, keeping it soaked in saliva as much as possible. No water, food or any kind of liquid was allowed during this collection period. The cotton wool was then placed inside the suspended container and closed with the lid. After collection, the Salivette with the cotton was kept under refrigeration until the Salivette tube with the cotton was centrifuged. After centrifugation, the cotton wool was removed and the saliva obtained was frozen at - 20° C.

4.1.5 **Determination of biochemical parameters**

The following methodologies were used to determine the biochemical parameters: UV enzymatic methodology to assess the level of cholinesterase, for creatinine, Labtest methodology, fixed time method with standard, for urea the UV enzymatic methodology, delta kinetic method with factor, for AST/TGO UV kinetic methodology - IFCC, delta kinetic method with factor, and for ALT/TGP UV kinetic methodology - IFCC, delta kinetic method with factor.

4.1.6 **Statistical analysis**

For the results obtained from the farmers' samples, the statistical analysis used was Student's *t-test* for variables with a normal distribution and Pearson's correlation coefficient, in order to obtain a relationship between the parameters analysed. The results are expressed as mean ± standard deviation. Differences will be considered significant when $p < 0.05$.

5 RESULTS

This study included 23 volunteers, who were sampled before the tobacco seedlings were planted in August 2014 and after the harvest in April 2015. The samples were taken from both men and women, 73.9 per cent of the 23 volunteers were men and 26.1 per cent were women. The age of the volunteers surveyed ranged from 20 to 60, but the largest number were in their 40s.

When asked how long these farmers have been planting and harvesting tobacco and consequently using pesticides, 39.13% reported using pesticides for approximately 21 to 25 years, 17.39% for 26 years, 8.69% for less than 20 years, 26.08% for 11 to 15 years and 8.69% for less than 5 years.

Of the 23 interviewees, 22 (95.6%) said they had already come into contact with pesticides during the preparation, planting and harvesting of tobacco, and only 1 said he had never had direct contact with a pesticide.

The frequency of pesticide use was a worrying issue, as 73.91% answered that they use pesticides 5 or more times a year, 13.1% use them 4 times a year, 8.7% use them up to 3 times a year and 4.39% didn't answer the question.

When asked about the use of PPE during pesticide application, 13.1 per cent of the interviewees said they used gloves, masks, glasses, boots and suitable clothing, 26.1 per cent only used gloves, masks, suitable clothing and boots, and the other 60.8 per cent used a maximum of three of the pieces of equipment indicated.

As for sanitising during or after pesticide application, 82.60% said they wash their hands before smoking or eating, while the other 13.1% don't sanitise, and 4.34% didn't answer.

Of those interviewed, 69.56% said that they observe the direction of the wind before applying pesticides, 26.08% apply the product without observing whether the wind is favourable or not and 4.34% did not answer.

The survey also sought to find out whether the farmers interviewed had ever had any kind of direct contact with pesticides, for example, contact with their skin, eyes or other parts of the body. When asked about this, 73.91 per cent of them said they had already had direct contact, 21.75 per cent had never had contact and 4.34 per cent did not answer.

When asked about the symptoms that appear after pesticide application, everyone mentioned at least one symptom, with 34.8 per cent citing eye irritation as the most common,

34.8% with muscle weakness, 39.13% with dizziness, 39.13% with malaise, 30.43% with salivation and the most prominent was headache symptoms with 60.86%.

Of the 23 farmers interviewed, 13.1 per cent (three) said they had already been hospitalised for pesticide poisoning, while the remaining 76.3 per cent had never had to be hospitalised for poisoning.

As for where the farmers buy the pesticides, 82.6 per cent said that they receive the pesticides used on their crops directly from the partner company, while the other 17.4 per cent buy them from agricultural companies.

According to figure 1, it can be seen that the salivary cholinesterase values after planting and harvesting the tobacco, the second collection, were significantly lower (P=0.0001) when compared to the samples from the first collection. The averages found were for the first collection (2959 U/L ± 58.0) and for the second collection (2284 U/L ± 25.1).

Figure 1: Evaluation of cholinesterase enzyme activity in saliva samples.

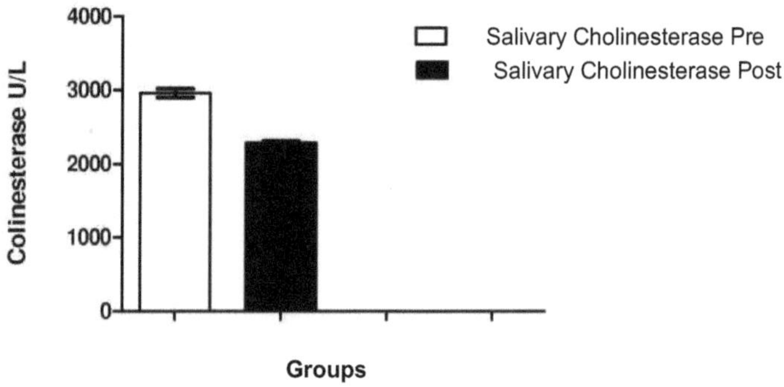

Data is expressed as mean ±SD. They were analysed using the student's t-test and values with P<0.05 were considered significant.

As can be seen in figures 2 and 3, there was no significant difference in the activity of the liver enzymes (TGO/AST and TGP/ALT). The averages found for the ALT enzyme were, for the first collection (26.3U/L ± 2.8DP) and for the second collection (24.2U/L± 2.9DP), resulting in a P=0.06. The averages found for the AST enzyme were, for the first collection (28.2U/P± 3.0DP) and for the second collection (23.0U/L±1.6DP), resulting in a P=0.01.

Figure 2: Evaluation of ALT enzymatic activity.

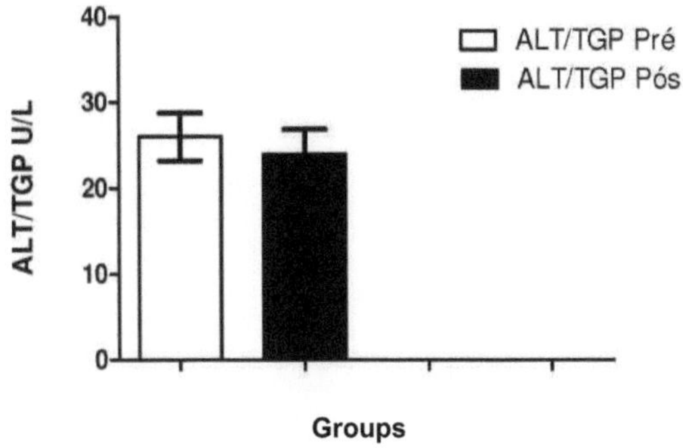

Data is expressed as mean ±SD. They were analysed using the student's t-test and values with P<0.05 were considered significant.

Figure 3: Evaluation of AST enzymatic activity.

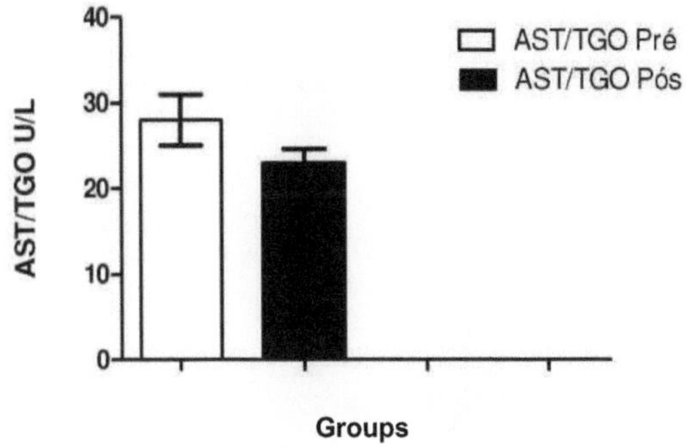

Data is expressed as mean ±SD. They were analysed using the student's t-test and values with P<0.05 were considered significant.

Figure 4 shows that there was no significance in the urea results between the first and second collections, with the averages found being (34.1mg/dl ± 1.6DP) and (33.2mg/dl ± 2.2DP) respectively, giving a P=0.07. In relation to creatinine, we observed a significant decrease in the post-planting period when compared to the pre-planting period (P=0.001). The

The average results found were for the first collection (1.2mg/dl± 1.6DP) and for the second collection (0.8mg/dl± 0.5DP).

Figure 4: Determination of serum urea concentration.

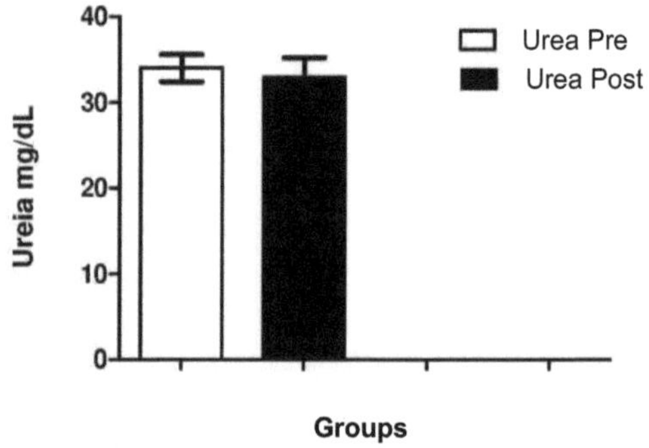

Data is expressed as mean ±SD. They were analysed using the student's t-test and values with P<0.05 were considered significant.

Figure 5: Determination of serum creatinine concentration.

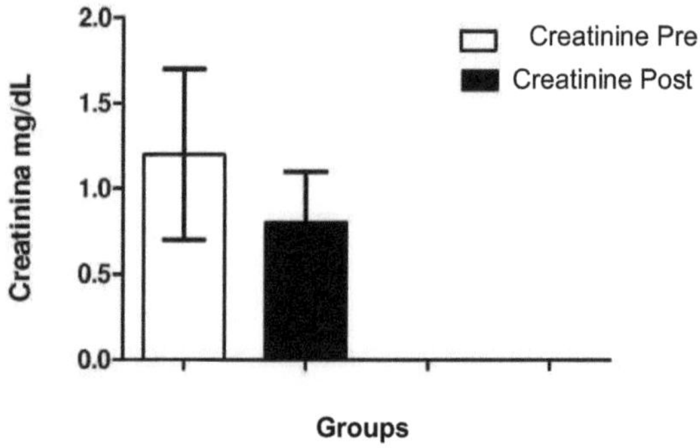

Data is expressed as mean ±SD. They were analysed using the student's t-test and values with P<0.05 were considered significant.

6 DISCUSSION

The research was carried out in the municipality of Paraíso - SC, and involved 23 volunteer farmers, from whom information and biological material were collected so that it could be carried out.

According to the data collected, it can be seen that both sexes come into contact with pesticides at all stages of tobacco production and are therefore exposed to pesticide poisoning.

It was observed that regardless of age, there is exposure to this risk, and that these people have been working with pesticides for a long time, which increases the risks to their health, in some cases exceeding 20 years of working with tobacco plantations. It should be emphasised that these farmers use pesticides several times during the period from planting to harvesting, which can be more than five times a year.

Only one of the 23 people surveyed had never had contact with pesticides. This is one of the data points that draws attention, as the contact rate is high at 95.6 per cent.

Among the population surveyed, 73.91 per cent had already come into contact with pesticides through their skin, eyes or other parts of the body. This is because most of the volunteers declared that they did not use all the PPE indicated for safety, which makes them more vulnerable to exposure. The vast majority, or 60.8%, use at most 3 of the 5 pieces of equipment indicated, which are gloves, suitable clothing, boots, masks and glasses.

According to Peres (2007) most farmers are unaware of the risks to which they are exposed and consequently neglect some basic occupational health and safety rules. According to Barros (2006), when PPE is not used when applying pesticides, the human body is exposed to toxic agents.

All the interviewees said they had experienced some symptoms after using the products, the most common being eye irritation, muscle weakness, dizziness, malaise, salivation and headache. However, 3 interviewees needed hospitalisation and medical treatment to recover, and were diagnosed with pesticide poisoning. The World Health Organisation (WHO) estimates that there are around three million pesticide poisonings and 220,000 deaths a year worldwide (PAHO/WHO, 1996).

The hepatic biomarkers AST and ALT did not show significantly different values between the first and second samples, showing that these patients do not have liver lesions that could be considered with these markers.

Analysing the renal biomarkers, the urea values remained within the normal range in both samples. The results found in the serum creatinine assessment showed a reduction in its levels in relation to the first and second collection, 1.2 mg/dl± 0.5 and 0.8 mg/dl± 0.5, respectively, despite this difference, the levels are still within the reference values for this test. GOETHEL, et al (2013), also found data similar to these in their study, in which they surveyed 40 tobacco farmers, but in only one stage, where the results of AST, ALT and urea were within the reference values, however, creatinine was high in most of the volunteers, characterising hypercreatinine, whose results were 1.28 mg/dL ± 0.34 , data consistent with our first collection stage where creatinine obtained results close to these levels 1.2 mg/dL ± 0.5.

By analysing the cholinesterase enzyme data in saliva samples collected from the volunteers, it was possible to achieve the results that were expected in the research. This proved that saliva is a satisfactory biological sample for diagnosing exposure to pesticides, as was already known for serum.

In the first stage of the research, the cholinesterase values in saliva samples averaged 2959U/L ± 58.0, remaining close to the reference value for serum cholinesterase, which is between 3200 and 9000 U/L. These results are in line with what we had hoped to achieve, given that the planting and handling of seedlings in the beds had already passed, where farmers are more exposed to pesticides, thus causing a pre-decrease in the marker.

In the period of the second collection, when the farmers were already in the tobacco cultivation and grading phase, they were in constant contact with the pesticides, where the cholinesterase levels found averaged **2284U/L ± 25.1**, values below the reference values, indicating signs of intoxication. When comparing the first and second samples, we saw a significant decrease in cholinesterase ($P<0.001$).

It is therefore possible to see a significant drop in cholinesterase levels in these farmers during the tobacco-growing period, i.e. there was contact with pesticides and consequent poisoning, confirming our assumption. However, there were cases in which there was no considerable variation in serum cholinesterase levels, when comparing the data from the first and second samples.

According to ARAÚJO, et al (2007), a study carried out in the city of Nova Friburgo in Rio de Janeiro, 102 farmers exposed to pesticides for 7 days prior to collection were studied in order to assess the values of Acetylcholinesterase (AChE) and butyryl-
cholinesterase -BuChE. The results were that erythrocyte cholinesterase (AChE) was reduced in 12.8 per cent of workers with signs suggestive of poisoning and in 3.7 per cent of cases without poisoning. Plasma cholinesterase (BuChE) was reduced in 17 per cent of cases of poisoning and in 25 per cent of cases without poisoning.

It is possible to relate the data from the research carried out to the information mentioned above, both of which were carried out with volunteers who had come into contact with pesticides. BuChE, which is an earlier marker of intoxication, showed a decrease because the farmers had recently come into contact with pesticides.

The data found in this study is considered to be consistent with the reality of the entire western region of Santa Catarina, as the practice adopted is similar in all farming families, and there is a lack of knowledge on the part of farmers about the importance of using safety equipment correctly in order to avoid contact and possible serious poisoning of their health.

7 FINAL CONSIDERATIONS

This study shows that the indiscriminate use of pesticides in the country is growing every year in order to improve and guarantee an increase in agricultural production. In this sense, it is necessary to take care and control those who apply the product, in order to avoid poisoning and damage to health.

The survey of 23 farmers in the municipality of Paraíso - SC, provided important information and data on the subject. It showed that a high number of people do not make proper use of safety equipment as a result of direct contact with pesticides, some of whom said they had already been poisoned while using the products.

Confirming this are the results of the cholinesterase tests on saliva samples, which are altered, indicating signs of pesticide poisoning. The use of saliva as a biological sample also proved reliable for this type of test. One of the positive points of tests like this is the practicality of collecting the material as it is a non-invasive method, thus facilitating studies and monitoring of the health of farmers in the western region of Santa Catarina and other regions.

It is a fact that most people are unaware of the real risks they are exposed to when they come into contact with these products. Because the initial symptoms of poisoning are common, such as malaise, weakness, headache, dizziness and others, they are not associated with pesticide poisoning and go unnoticed, further aggravating the problem.

This leads us to conclude that there is an urgent need for more work to raise awareness of the importance of correct handling and the use of PPE, in order to increase safety and quality of life for farmers.

REFERENCES

ALMEIDA, Eva Aparecida et al. **Agrotoxics and the risk to health among tobacco farmers.** Jul/Dec 2011. Available at: <www.revista2uepg.br>. Accessed on: 29 Oct 2014.

ALMEIDA, Pedro José. **Poisoning by pesticides**. 2002.

AMORIN, L. C. A. **Biomarkers and their application in the evaluation of exposure to environmental chemical agents**. Rev. Bras. Epidemiol. 2003.

ANVISA. National Health Surveillance Agency. **Pesticides and Toxicology**. Available at:< www.anvisa.gov.br>. Accessed on: 09 Apr 2015.

ARAÚJO, Alberto José de et al. **Multiple exposure to pesticides and health effects: cross-sectional study in a sample of 102 rural workers.** Ciência & Saúde Coletiva. 2007. Nova Friburgo, RJ.

BARBOSA, Anderson de Almeida; et al. Profile of aspartate aminotransferase and alanine aminotransferase and liver biometry of Japanese quails. **R. Bras. Zootec.**, v.39, n.2, p.308-312, 2010.

BARROS, D.C. **Occupational exposure to carbamates and organophosphates.** Niterói. UFF. 2006.

BRAZIL. **OSWALDO CRUZ FOUNDATION**. NATIONAL TOXIC-PHARMACOLOGICAL INFORMATION SYSTEM. Annual statistics of intoxication and poisoning cases. Rio de Janeiro, 1993. 74p.

CAZARIN, Gisele et al. Hematological diseases and environmental risk: the importance of clinical records to epidemiological surveillance. **Revista brasileira de epidemiologia.** São Paulo, v. 10, n. 3, 2007.

CHAMBERS, H.W.; BOONE, J.S.; CARR, R.L.; CHAMBERS, J.E. **Chemistry of organophosphorus insecticides.** In Krieger R. Handbook of pesticides toxicology: Agents 2nd edition. California: Academic press. 2001.

CHESTER, Douglas. **Currently available salivary diagnostics.** Nov.2005. Available at: <www.colgateprofissional.com.br>. Accessed on: 19 Sep.2014.

FARIA, N.M.X.; ROSA, J.A.R.; FACCHINI, L.A. **Pesticide poisoning among rural fruit**

workers. Bento Gonçalves. RS. Public Health Code. 2009.

FERNANDES, Daniel et al. **Organophosphate poisoning.** Volumen 18, No. 1 - Enero - Junio de 2010.

FUKUTO, T.R. **Mechanism of action of organophosphorus and carbamate insecticides**. Environ Health Perspect. 1990.

LARINI, Lorival. **Toxicology of pesticides.** São Paulo: 1ª Ed. 1999.230p.

LOTTI, M. **Clinical toxicology of anticholinesterase agents in humans.** In: Krieger R. Handbook of pesticides toxicology. Agents 2nd edition. California Academic press. 2001.

MARONI, M.; COLOSI, C.; FERIOLI, A.; FAIT, A. **Organophosphorus pesticides.** Toxicology. 2000.

MÍDIO, Antonio Flávio; SILVA, Erasmo Soares da. **Organophosphate and carbamate insecticides-acaricides**. São Paulo: Roca, 1995. 84p.

NUNES, Pedro Pimentel; Adelino Leite Moreira. **Hepatic physiology.** Faculty of Medicine of the University of Porto Physiology Department. Porto, Academic Year 2006 / 2007.

PAHO/WHO. **Manual on health surveillance of populations exposed to pesticides**. Brasilia: PAHO/WHO, 1996.

PASIANI, Juliana de Oliveira. **Knowledge, attitudes and practices of rural workers in relation to the use of pesticides and biomonitoring of exposure.** Brasilia. DF. 2012.

PERES, F., et al. **The impacts of pesticides on human health and the environment.** Rev. Ciência e Saúde Coletiva. V 12. N 1. RJ. 2007.

PINTO, Samanta B. **Comparison between AST (Asparto Aminotranferase) and ALT (Alanine Aminotranferase) levels in the presence and absence of peridoxal phosphate.** Porto Alegre. RS. Jun. 2010.

PRONAF. **National Programme to Strengthen Family Farming.** Available at: <www.pronaf.gov.br>. Accessed on: 01 Apr 2015.

RANG, H.P., DALE, M.M.; RITTER, J.M. MOORE, P.K. **Cholinergic transmission**. Pharmacology. 5th Edition. São Paulo. Elsevier Publishing House. 2004.

REBELO, M.; VASCONCELOS, R. A., BUYS, B. D. M. C.; REZENDE, J.A.; MORAES, K.O.C.; OLIVEIRA, R.P. **Pesticides and related products commercialised in Brazil in 2009.** An environmental approach. Brasilia. IBAMA. 2010.

REINER, E.; SIMEON-RUDOLF, V. **Methods for measuring cholinesterase activities in human blood. Toxicology of organophosphate and carbamate.** 1st Edition. San Diego. Elsevier Academic press. 2006.

RIBEIRO, Amanda C. C.; MELLA, Eliane A. C. **Occupational organophosphate poisoning** - The importance of cholinesterase dosage. Jul./Dec.2007. Available at: <www.cesumar.br/pesquisa/periodicos>. Accessed on: 02 September 2014.

SATOH, T. **Global epidemiology of organophosphate and carbamate poisonings.** In Grupta RC. Toxicology of organophosphates and carbamates. 1st Edition. San Diego. Elsevier Academic press. 2006.

SCHUTZEMBERGER, Monica E. et AL. **Biochemical analysis of salivary fluid from individuals with periodontal disease.** 2007. Available at: <WWW.uel.br/pessoal/buzato/pages/arquivos/seminarios/a6>. Accessed on: 15 Aug 2014.
SILVA, Jandira Maciel da et al. **Pesticides and work: a dangerous combination for the Brazilian agricultural workers health**. Ciência e saúde coletiva . vol. 10. 2005.

SINDAG. **National Union of the Agricultural Defence Products Industry.** Available at: <www.sindag.com.br>. Accessed on: 02 May 2015.

SINITOX. **National Toxic-Pharmacological Information System.** Available at: <www.fiocruz.br/sinitox>. Accessed on: 20 Apr 2015.

STROPARO, Luciana de F. et al. **Evaluation of exposure to organophosphates among tobacco growers in a locality in the municipality of Irati - Paraná.** Jan./Jun.2011. Available at: <http://.www.revistas2.vepg.br/index/php.exatas>. Accessed on: 08 Aug.2014.

THOMPSON, H.M. **Esterases as markers of exposure toorganophosphates and carbamates.** Ecotoxicology. 1999.

TRÁPE, Angelo Zanaga. **Toxicological effects and records of pesticide poisoning.** Available at : <www.tudosobretomate.com.br>. Accessed on: 29 October 2014.

TROIAN, Alessandra et al. **The use of pesticides in tobacco production:** Some perceptions of farmers in the Candido Brum community, in the municipality of Arvorezinha (RS). Jul.2009. Available at: <www.sober.org.br/palestra/13/844.pdf>. Accessed on: 21 August 2014.

WHO. **World Health Organisation. Organophosphorus insecticides: A general introduction.** Environmental health criteria. 1986. Available at: <www.wpro.int/NR/rdonlyres>. Accessed on: 28 March 2015.

WHO, **World Health Organisation.** Environmental health criteria 155, Biomarkers and risk assess net: Concepts and principles. 1993. Available at: <www.inchem.org/documents>. Accessed on: 10 Apr 2015.

ANNEXES

ANNEX A - Informed Consent Form

INFORMED CONSENT FORM

You are hereby invited to participate as a volunteer in the following research project: **CHOLINESTERASE EVALUATION FOR THE DETECTION OF ORGANOPHOSPHORATES AND CARBAMATES IN TOBACCO GROWERS IN THE CITY OF PARAISO, SANTA CATARINA.**

JUSTIFICATION, OBJECTIVES AND PROCEDURES: The reason for proposing this study is to obtain data on the rates of Organophosphate and Carbamate poisoning that have been harming a large number of people. The research is justified by the fact that it is extremely important to carry out studies that seek a better understanding of the health of tobacco farmers in the city of Paraiso, in relation to the use of pesticides on crops. The aim of this project is to investigate an enzyme called cholinesterase in both blood (serum) and saliva, as well as a full blood count to assess infections, inflammation and anaemia. The data collection procedure(s) will be as follows: After 8 hours of fasting, blood will be collected using a 25/7 needle and a 10 mL syringe, after the arm has been tourniqueted and cleaned with 70% alcohol. The blood sample will be placed in collection tubes with anticoagulants (EDTA and heparin) and without anticoagulants for processing. At the end of the collection, a bandage will be placed on the site and it will be checked that everything is OK with the participant, before the participant is released. For the saliva collection, the participant will be asked to rinse their mouth with water three times after an 8-hour fast, and then provided with a Salivette® cotton swab. Participants will be instructed to remove the cotton wool and place it in their mouth for 3 minutes. The cotton wool will then be placed back in the container for centrifugation.

DISCOMFORTS, RISKS AND BENEFITS: Your participation in this study may generate some kind of discomfort regarding blood collection such as slight pain and discomfort at the time of puncture, and in difficult collections you will have the risk of getting a haematoma due to blood leakage, you may also feel embarrassed at the time of collecting saliva, however the benefit that will be obtained, will be important for both the participants and the farming population in general, because through this study we will be able to collect biochemical, inflammatory, toxicological and statistical parameters that could be useful in the diagnosis and monitoring of comorbidities related to pesticide poisoning. The participants will have their results analysed by the researchers and will subsequently be provided with important information about their metabolism and the possible clinical complications that the excess of organophosphates and carbamates is producing in their bodies, so that the benefits outweigh the risks.

FORM OF MONITORING AND ASSISTANCE: The researchers will analyse the results of each participant and then pass on the information individually. Participants who do not present altered results will be advised that they do not have any complications regarding the analyses carried out and will receive health information to continue in this situation. For participants who show alterations in any of the results assessed, they will be given individual guidance on the complications that could occur in their metabolism and possible diseases that could be triggered. They will then be advised to go to a specialist doctor, who will be in charge of the participant. The researchers are responsible for providing copies of all the dosages taken so that they can be taken to the clinician. The researchers will also provide their private telephone numbers so that the doctor can contact them and answer any questions that may arise.

clarify any doubts that may arise. The researchers will advise these participants on healthy lifestyle habits and attitudes that will help to improve or stabilise the tests.

GUARANTEE OF DISCLOSURE, FREEDOM OF REFUSAL AND GUARANTEE OF SECILITY: You may request clarification about the research at any stage of the study. You are free to refuse to take part, withdraw your consent or stop taking part in the research at any time, whether for reasons of embarrassment or otherwise. Your participation is voluntary and refusal to participate will not result in any penalty or loss of benefits. The researcher(s) will treat your identity with professional standards of confidentiality. The results of the laboratory and anthropometric tests will be given to you and will remain confidential. Your name or material indicating your participation will not be released without your permission. You will not be identified in any publication that may result from this study. This consent form is printed and signed in two copies, one copy will be given to you and the other will be kept by the researcher(s) responsible.

PARTICIPATION COSTS, REIMBURSEMENT AND INDEMNIFICATION: Participation in the study will not entail any costs for you and no compensation will be provided. In the event that you suffer any damage as a result of this research, the participants undertake to guide, pay for and accompany the participant until the damage is resolved.

STATEMENT BY THE PARTICIPATING SUBJECT OR THE PERSON RESPONSIBLE FOR THE PARTICIPATING SUBJECT:

I .... was

I have been informed of the objectives of the above research in a clear and detailed manner and have clarified my doubts. I know that at any time I can request further information and/or withdraw my consent. Those responsible for the above research have assured me that all my data will be kept confidential. If I have any questions, I can call the researcher in charge, Eduardo Ottobelli Chielle, who lives at Rua da Fonte 175 Jardim Acácias, São Miguel do Oeste - SC (49) 3631 1072 / 1084 or contact the Human Research Ethics Committee at Unoesc and Hust, Rua Getúlio Vargas, n° 2125, Bairro Flôr da Serra, 89600-000- Joaçaba - SC, Phone: 49-3551-2012. I declare that I agree to take part in this study. I have received a copy of this informed consent form and have been given the opportunity to read it and clarify my doubts.

Signature of the research subject or **fingerprint**.

Signature:

Name legible:

Address:

RG

Phone:

__________ Date//_

Signature of the researcher responsible

__________ Date//_

APPENDICES

APPENDIX A - QUESTIONNAIRE APPLIED

1. Name:

2. Sex:
   ( ) Male ( ) Female

3. Age: ______________________.

4. How many years have you been a tobacco grower?

5. Do you come into contact with pesticides during the preparation, planting and harvesting of tobacco?

   ( ) Yes ( ) No

6. How often do you use pesticides?

   ( ) 1 time per year ( ) 3 times per year ( ) 5 times per year

   ( ) 2 times a year ( ) 4 times a year ( ) More than 5 times a year

7. How long do you use pesticides?

   ( ) Less than 5 years old ( ) 8 to 9 years old

   ( ) From 6 to 7 years ( ) More than 10 years

8. Do you use any of this safety equipment when applying pesticides?

   ( ) Gloves ( ) Masks ( ) Glasses ( ) Suitable clothing ( ) Boots.

9. Proper hand hygiene before smoking or eating, during or after application
   ( ) Yes ( ) No.

10. Do you check the wind direction before applying the pesticide?

    ( ) Yes ( ) No

11. Have you ever spilt the pesticide on your skin, eyes or any other part of your body when handling

it?

( ) Yes ( ) No

12. Have you experienced any of these symptoms after applying the pesticide?

( ) Blurred vision ( ) Eye irritation ( ) Salivation

( ) Dizziness ( ) Tearfulness ( ) Coughing

( ) Headache ( ) Nausea/vomiting ( ) Malaise

( ) Sweating ( ) Muscle weakness ( ) Tachycardia

( ) Irritability ( ) Skin irritation

( ) Abdominal pain ( ) Shortness of breath

( ) Other symptoms ____

13. Have you ever been hospitalised for poisoning?

( ) Yes () No

14. Where do you buy pesticides?

( ) Farming ( ) Direct with tobacco company

( ) Veterinary ( ) Argentina

( ) Other. ____

MIX
Papier aus verantwortungsvollen Quellen
Paper from responsible sources
FSC® C105338

Printed by Books on Demand GmbH, Norderstedt / Germany